The Wet Basement Manual:
Methods and Innovations Employed Since 1947

SECOND EDITION

The Aberdeen Group
A division of Hanley-Wood, Inc.
426 S. Westgate St.
Addison, IL 60101

© 1999, 1993 by The Aberdeen Group
All rights reserved. Except as permitted under the United States Copyright Act of 1976, no part of this work may be reproduced or distributed in any form or by any means or stored in a database or retrieval system without permission of the publisher.

Printed in the United States of America

09 08 07 06 05 04 03 02 01 00 99   5 4 3 2 1

Editor: Carolyn Schierhorn
Technical Reviewer: Bruce Suprenant
First Edition Editor: Midge Anderson
Art Director: Joan E. Moran
Publisher: Mark DiCicco

The information contained in this work has been obtained by The Aberdeen Group from sources believed to be reliable. However, neither The Aberdeen Group nor its authors guarantees the accuracy or completeness of any information published herein, and neither The Aberdeen Group nor its authors shall be responsible for any errors, omissions, or damages arising out of the use of this information. This work is published with the understanding that The Aberdeen Group and its authors are supplying information but are not attempting to render engineering or other professional services. If such services are required, the assistance of an appropriate professional should be sought.

Library of Congress Cataloging-in-Publication Data

Maurice, A. E.
   The wet basement manual : methods and innovations employed since
1947 / by A.E. Maurice : architectural drawings by Raymond T.
Guertin. – 2nd ed.
     p   cm.
   ISBN 0-924659-50-5
   1 Waterproofing. 2. Dampness in basements.   I. Title.
TH9301.M38   1999
   693.8'92—dc21   99-24726
                                                                                CIP

Item No. 1060

The Wet Basement Manual:
Methods and Innovations Employed Since 1947

SECOND EDITION

by
A.E. Maurice
Danvers, Massachusetts

Architectural drawings by
Raymond T. Guertin, Architect

The Aberdeen Group®
a division of Hanley-Wood, Inc.

426 S. Westgate St., Addison, Illinois 60101
Telephone: 630-543-0870 Fax: 630.543.3112

CONTENTS

Author's Preface 1
The French Drain System 3
Constructing the Sump Pump Holding Chamber 8
A Finished Basement 16
Leaking Basement Windows 18
A House With a Stone Foundation 21
Installing a Drain System Outside of
  Foundation Walls 23
A Concrete Block Foundation Wall 25
Damp-Proof Below-Grade Walls 25
Hydrostatic Pressure Under a Concrete
  Floor Where Head Room is Available 27
Hydrostatic Pressure Under a Concrete
  Floor With Sandy Soil and No Head Room 29
A Single Sump Pump Installation 29
A House Built on Wet Land 32
A House Built on a Downgrade Slope 32
Establish House Grade 34
Installing Drains in Front of Garage Doors 35
A Sump Pump Lift Chamber 39
A Dry Well Chamber 41
A House Built on a Concrete Slab 41
Outside Sump Pump Chamber 43
Controlling Downspout Water 45
Using a Blow Pipe to Excavate
  Under or Around Obstructions 45
Ledge Rock in Basement 49
A House Foundation Wall with No Footing 53
Supplementary Information 56

ILLUSTRATIONS

Figure 1. Contact Points 2
      2. Contact Points–Cracked Wall 5
      3. French Drain With Holding Chambers 7
      4. Dual Sump Pumps 9
      5. Sump Pump Holding Chamber 11
      6. Sump Pump Holding Chambers
         (side view) 13
      7. High Foundation Footing 15
      8. French Drain System With Vapor Barrier.17
      9. Window Drain..19
    10. Another Way to Drain a
         Basement Areaway...................................20
    11. Drain Inside a Stone Foundation 22
    12. Outside Drain System—
         Concrete Block Foundation 24
    13. Damp-Proof Foundation Walls 26
    14. Hydrostatic Pressure—
         New Floor Poured Over Old....................28
    15. Sump Pump Installation...........................30
    16. Single Sump Pump Chamber 31
    17. The Outside System—
         Concrete Foundation 33
    18. Driveway Sloping Down to Garage 36
    19. New Grated Channel Slope System........37
    20. Channel Slope System With
         Optional Mini Catch Basin 38
    21. Sump Pump Lift Chamber 40
    22. Dry Well Chamber 42
    23. Outside Drain System—House on Slab....44
    24. Downspout Water Diversion 46
    25. Downspout Extension 47

26. Downspout Extension to Dry Well 48
27. The Blow Pipe 50
28. Blow Pipe Excavation Under Tank
    or Other Type of Obstruction..................51
29. Excavation Around Tank
    or Other Obstruction 52
30. Exposed Ledge Rock 54
31. French Drain for Footingless Wall..........55
32. Ferric Oxide Removal System................61
33. Water and Radon Removal 63
34. Radon Gas Removal System—
    Dry Basement 64
35. Water and Radon Removal (when a sump is
    installed in an outside concrete chamber) ...65
36. Water and Radon Removal—
    Outside System 66

Author's Preface

In 1923, Ernest Maurice, my father, came to the United States from Canada and started working for a general contractor, where he gained experience in waterproofing. In 1933, Ernest started his own contracting business. As his four sons became old enough to work, they joined the family business. My oldest brother, Edmond Maurice, worked in the waterproofing business for 50 years before retiring. Many times I have called on him for advice.

After coming home from the service during World War II, I went back to work in my father's business. Shortly thereafter, like my father, I started my own construction company. This book contains many of the insights I have acquired over the years about the waterproofing business. I hope it proves useful to contractors and do-it-yourselfers alike.

Although the book begins with instructions for constructing a French drain system, it assumes that more easily corrected water problems—such as plumbing leaks, improperly draining downspouts, and foundation wall cracks—have been eliminated as causes of trouble. Several of the most common water problems, requiring less extensive solutions than the French drain system, are discussed in the Supplementary Information section at the back of the book.

Disclaimer

The information in this book is correct to the best of my knowledge. The recommendations or suggestions herein are without guarantee since actual conditions in the field vary from situation to situation. This book should be used in conjunction with sound judgment by those familiar with other recognized construction techniques, as an additional source of information, and to supplement current knowledge on this topic.

It is a short book, meant to be read completely before the reader begins work. Information contained in early chapters is not necessarily repeated later, even though it may be relevant. No warranty is expressed or implied; and in no event shall the author be liable for any loss or damage whatsoever, including but not limited to direct, indirect, incidental, or consequential damages.

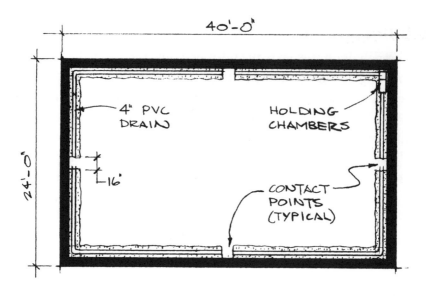

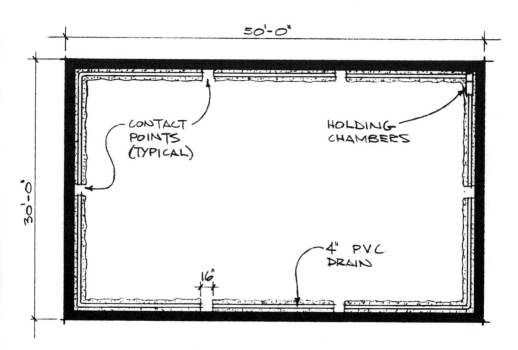

Figure 1: Contact Points

THE FRENCH DRAIN SYSTEM

The French drain system is the most practical method of correcting a water problem in a wet basement. Such a system is low-tech by modern standards and, in many cases, can be installed fairly quickly. In any waterproofing project, it is important to complete the job as quickly as possible, especially when heavy rain or other sources of water damage are threatening.

A French drain is basically an underground passage made by filling a trench with loose stones or rubble. For more than 55 years, the Maurice family has successfully adapted French drains to solve a variety of basement water problems.

Three pieces of information are needed in order to assess any wet basement problem:
- The thickness of the basement floor
- The type of soil under the floor
- The height of the water table

A test hole will help you determine the facts. Cut a test hole in the floor with a stone drill or small jackhammer at the location where a sump pump holding chamber will be installed. In locating the sump pump, the most important consideration is to accommodate discharge piping outside the building.

Dig a test hole

The test hole procedure is worth the time involved. If the soil is sandy and wet, it is best to repair your test hole and wait for a drier season. To fill in a sizable hole (about 12x12 inches), mix two parts sand to one part cement; then put together a primer "paste" of cement and water and trowel it on the sides of the hole before filling it with the two-to-one patch material. If your test hole is small, it can be filled with a water plug or hydraulic cement.

Contact points

If your test hole reveals dry soil, proceed by cutting the concrete floor with a concrete saw or a 75- to 85-pound jackhammer near the foundation walls that show water seepage. If your floor is thin and you are able to do the work, a sledgehammer can sometimes do the job.

Cut the concrete floor, measuring from the foundation wall onto the basement floor approximately 14 inches, or 16 inches if the footing is in the way. It is important to leave uncut sections of floor every 15 to 20 feet, to serve as wall supports for safety purposes. These uncut sections are known as contact points, and they are approximately 16 inches long. Where wall cracks are evident, leave two wall supports on either side of the crack and cut out 48 inches between them (*see Figures 1 and 2*).

After the concrete floor trench has been cut out and the concrete removed, deepen the trench to 12 inches from the top of the concrete floor. Remove small quantities of dirt with pails; a conveyor belt may be economical for big jobs. If an excess of water flows into the basement after digging, install a temporary sump pump in the area where permanent sump pumps will be installed.

Sump pumps

The trench you have created will cause water to enter at a lower level, resulting in more water entering into the sump pump chamber or chambers. Through the years, I have become a strong believer in installing two sump pumps side by side. In addition, a gasoline-powered auxiliary generator is an essential investment for any building where flooding and power outages have occurred. I also recommend that you manually activate your sump pumps at least once a year to make sure they are in working order.

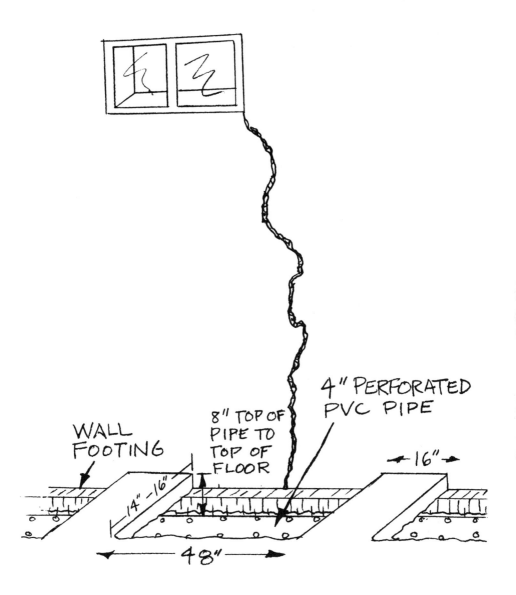

Figure 2: Contact Points—Cracked Wall

Your estimate of the amount of water coming in will determine the size of the chambers and the capacity of the pumps to be installed. When locating the pumps, consider where the sump pump water will be discharged. The water must flow away from the house on a downgrade slope, using plastic piping that will subsequently be laid. More information on chambers and pumps is provided later in this book.

Laying PVC pipe

When the digging is complete, including removal of the dirt under the contact points, start laying 4-inch perforated plastic pipe, with the holes up. Begin from the sump pump chambers and continue around the basement, using elbows in the corners and returning to the other side of the pump (*see Figure 3*). Start filling the trench with clean crushed stone, from ¾ inch to 1½ inches in size, to a height enabling the replacement concrete floor to be level with the existing floor and of equal thickness. The top edge of the sump pump holding chamber should be lower than the existing concrete floor so that a ¾-inch plywood cover installed over the chamber can be flush with the existing floor. This cover will prevent concrete from entering the chamber. After the concrete is poured, be sure that the cover has not bonded to the concrete and can be removed.

Preparing for new concrete

The abutting concrete should have a clean surface to bond to. The cut concrete floor edge can be washed with a heavy spray and brushed before the new concrete trench is poured. Then cover the stone with concrete, flush to the existing floor.

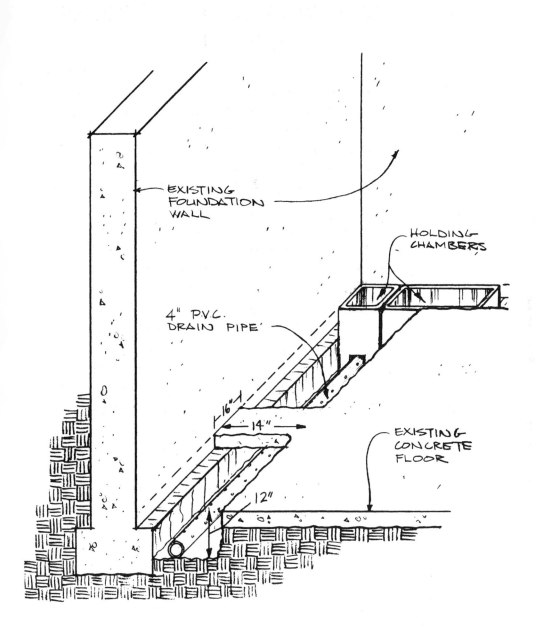

Figure 3: French Drain with Holding Chambers

CONSTRUCTING THE SUMP PUMP HOLDING CHAMBER

Sump pump holding chamber

Clay chimney flue liners, which can be purchased in a variety of sizes, make good holding chambers for sump pumps. A flue liner 12x12 inches and 24 inches in depth can hold a dewatering submersible pump that will pump 60 gallons of water per minute and can lift the water to 25 feet in height. It operates on a single-phase alternating current at 115 or 230 volts; the discharge opening of the pump is 1½ inches.

A larger chamber of 12x18 inches, 24 inches in depth, will hold a dewatering submersible pump that will pump 110 gallons per minute and can raise water 50 feet in height. This larger pump, placed in the larger chamber, will also operate on a single phase at 115 or 230 volts; its discharge opening is 2 inches. (In my own experience, the Hydromatic OSP33 and SPD50H have been the most dependable, efficient, long-lasting, and trouble-free sump pumps. However, at least 12 other pumps currently on the market can be used for this purpose.) *Figure 4* illustrates a two-pump arrangement.

Installing the chambers

For a two-pump system, dig to a depth of approximately 27 inches from the top of the concrete floor to the bottom of the sump pump hole, and place approximately 2 inches of crushed stone inside the hole. The length and width of the hole depend on the size of the holding chambers. Place the flue liners into the hole, seating them firmly onto the stone. To make bases for the pumps, place a piece of 1½- or 1¾-inch-thick precast patio block inside each chamber. The patio block slabs will prevent stones from entering the pump impellers and stopping the pumps' action. The imperfect fit of the prefabricated slabs inside the chambers

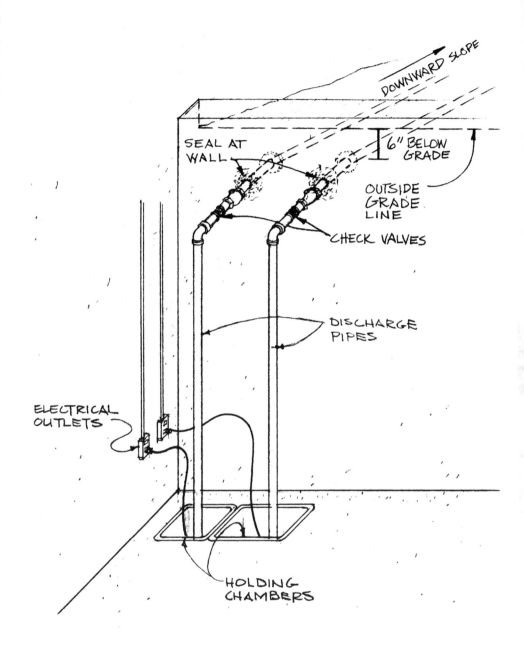

Figure 4: Dual Sump Pumps

will allow water to enter or drain out of the chambers.

If you are installing two pumps of different sizes, elevate the larger pump on two layers of patio block. The higher pump will work less often, leaving removal of smaller amounts of water to the smaller pump.

After the holding chambers are installed, fill around them with stone to the bottom of the existing drain trench. After laying the PVC drain pipe, continue filling with stone to the bottom of the floor slab.

Cut holes in the chambers

Before installing the sump pump holding chambers, cut a square hole 4x4 inches in two sides of each holding chamber (*see Figure 5*). This is achieved by using a power saw with a concrete cutting blade. Measure down from the top of the holding chamber 11 inches to the bottom of the hole. When two holding chambers are used, two holes should be cut at the same level so that water can enter from one chamber into the other. In this way either pump will help dissipate excessive water or kick in if the other should fail. Each sump pump has a check valve to prevent the water from backflowing out of the discharge pipes. It is important to install a separate electrical outlet and circuit breaker for each pump and for the gasoline auxiliary generator previously mentioned. [Caution: (1) All installation of wiring must be done by a qualified, licensed electrician. (2) All pertinent electrical codes should be followed.]

Two pumps

I strongly recommend the two-pump system as a precaution against any further basement flooding. (In a few extreme cases, I have installed three pumps in one basement.) Depending upon the circumstances, any combination of pumps could be installed. The smaller pump can withstand the wear of more frequent operation, and it should be installed so that the depth of

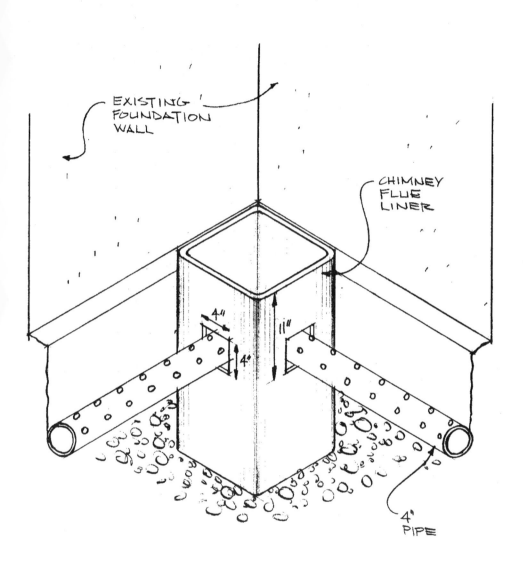

Figure 5: Sump Pump Holding Chamber

water before the pump starts working is 2 inches lower than the depth of water that activates the larger pump (*see Figure 6*).

Placing the larger pump on two layers of precast concrete patio block and putting the smaller pump on one layer of the concrete accomplishes this difference in activation very simply. Also, your sump pump installation instructions will include information about setting the activation point of your pump. With this two-pump system, excessive water will be taken care of by the larger backup pump. And if one of the two pumps should fail, repairs can be made on the defective pump while the working pump continues to function properly.

Discharge lines

Once the size of the dewatering submersible pumps has been determined and the appropriate chambers have been installed, discharge piping must be laid. Two pumps can usually tie into one 2-inch discharge line; but if there is a great deal of water, you should provide a discharge line for each pump. Drill a hole through the foundation wall 6 inches below the outside grade, and install 2-inch solid PVC piping through the foundation and along the base of an outside trench, keeping it on a downward slope (*see Figure 4*). Normally, discharge pipes are 4 to 10 feet long. If possible, end the pipe with an angled cut to match the slope of the yard or soil.

The piping in *Figure 6* shows that either pump can be removed for repair. If the pipe slips out of joint, the check valves prevent water from returning into the sump pump chamber, allowing the other pump to keep working.

Be ready for more rain

Always start a French drain system at the location of the sump pump hole or at the gravity flow outlet. These measures are important while the work is in progress because, if an unexpected rainstorm should

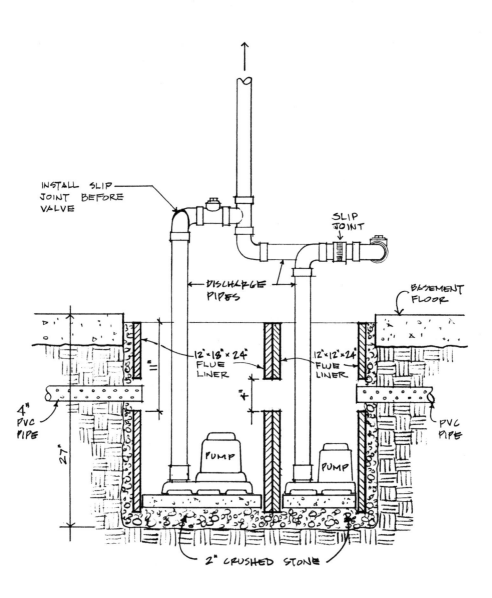

Figure 6: Sump Pump Chambers (side view)

occur, it would still be possible to install a pump that would keep the water table down. Where a sandy subgrade exists, the installer could excavate and install one section of pipe in the trench and cover it with stone to stop erosion caused by water shifting the sand.

If the situation involves a natural gravity flow system (which is just a natural downhill flow of water away from the house, unaided by a pump), dig a trench through the foundation wall or under the wall footings for holding the discharge pipe. The discharge pipe will connect with those laid on the downgrade slope.

High footing

If while installing the French drain system, you find that the footing under the basement floor is high (not leaving enough room to lay 2 inches of crushed stone over the foundation and under the concrete floor), cover the top of the footing with ⅜ inch of crushed stone and pour a thin slab of concrete over the footing. This procedure allows the water that flows between the footing and the concrete floor to enter into the French drain system (*see Figure 7*). If the concrete footing should be too high, cut the corner of the footing with a jackhammer blade as needed.

Chimney or fireplace base

The French drain system can be adapted to help basements that show evidence of water entering around either a chimney base or a fireplace base. One sign may be efflorescence (deposits of salt crystals) on the fireplace base due to the evaporation of water. Because of capillary attraction, water will be drawn to the surface around the fireplace base, resulting in moisture on the basement floor. A solid chimney base or Lally column footings that are flush with the basement floor and much thicker than the floor will sometimes cause capillary attraction. Because the massive footings are deep, water moves into their outer and upper surfaces.

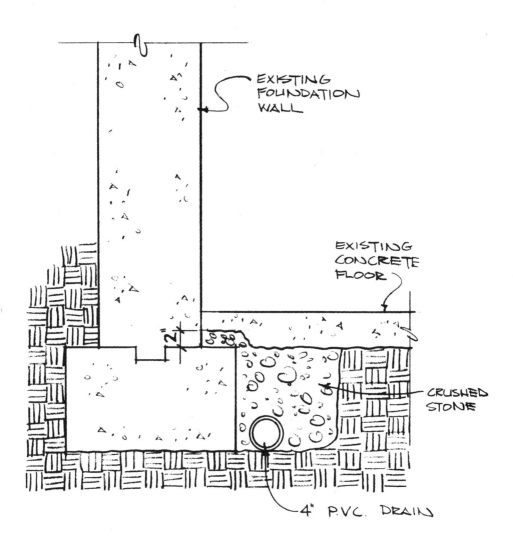

Figure 7: High Foundation Footing

Start by cutting the concrete floor approximately 12 inches in width around the leaking chimney or fireplace base. Dig down 12 inches and cut a trench to the leaking base, which should be the same depth as the existing perimeter drain. The system under construction should be drained. Lay pipe at the base of the trench connecting into the French drain system. Then, lay stone over the pipe and pour concrete over the stone.

A FINISHED BASEMENT

Extra care required

When installing a French drain system in a finished basement with wall paneling or plasterboard-finished walls, cut a trench only on the sides that have the water problem. You must cut the concrete floor under the finished wall back to the foundation wall and remove the soil and concrete there. Next, dig the sump pump hole and lay the perforated pipe. Afterwards, fill the trench with crushed stone over the pipe and backfill with crushed stone to the foundation wall. It is imperative to keep the concrete approximately 1 inch away from the foundation wall. This space or weep hole is necessary to allow water to enter the stone trench from any leaking wall ties, soil pipes, or cracks in the foundation walls.

Foundation-wall cracks

If conditions involve a high water table and fast water recovery with sandy soil, omit the weep space behind the finished panel wall. Unlike the above weep space system, be sure to push the concrete under the plate to make a seal against the foundation wall. If the foundation wall leaks, this can be remedied by digging on the outside next to the foundation wall, cleaning the cracked surface area, and drying with a torch, if necessary, 6 inches on each side of the crack. Then apply a plastic roofing cement ½ inch thick to cover

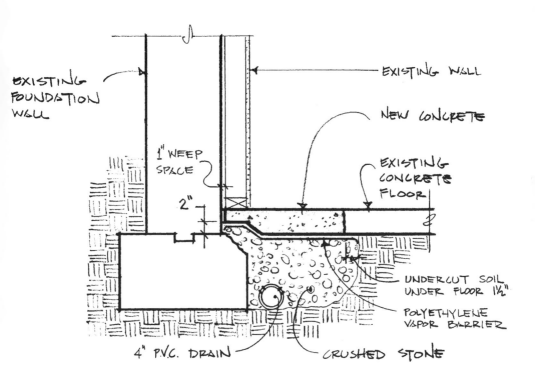

Figure 8: French Drain System with Vapor Barrier

the prepared surface. Protect the plastic cement from disturbance by covering it, from bottom to top, with a 6-mil sheet of polyethylene (*see Figure 8*). Be sure to secure the polyethylene sheet with a strapping nailed to the house.

LEAKING BASEMENT WINDOWS

Another condition that requires water control is where water enters through or around windows. There are two probable causes: (1) the foundation wall is too low, or (2) the grading is inappropriate. Either way, the window is now situated below the existing grade. Efforts to divert the water by installing an areaway or window well may only attract and collect water if the drainage is not corrected, too.

Window well or areaway

The solution is to dig outside the house foundation around the full width of the window area. Dig to a depth of 18 to 24 inches below the bottom of the window. Drill a hole from the inside of the basement (centered under the window) 2 inches up from the bottom of the window well through to the outside wall. Install a 1½-inch plastic pipe through the hole and cement the pipe on each side of the foundation wall.

Fill the areaway or window well with clean crushed stone to a level of 4 to 5 inches below the bottom of the window opening. Typically, a galvanized steel liner is installed; backfill around the outside of this unit with soil. The drain pipe will connect with the French drain system under construction in the basement. If a French drain system is not being installed, the water will have to be piped into a sump pump hole located in the basement floor (*see Figure 9*).

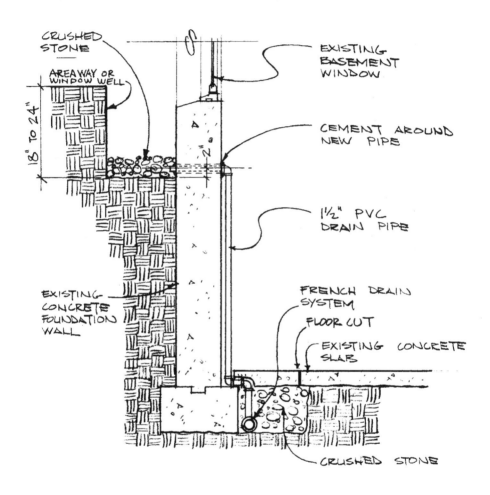

Figure 9: Window Drain

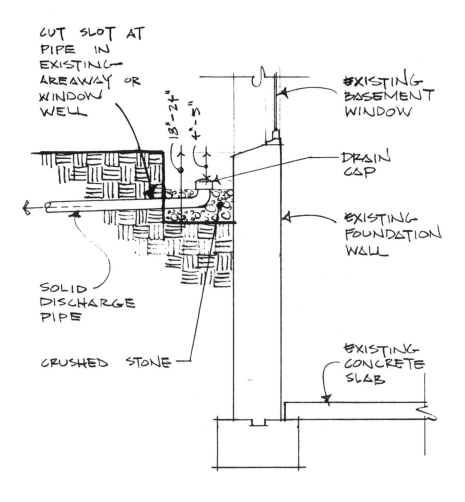

Figure 10: Another Way to Drain a Basement Areaway

If gravity flow will permit drainage away from the building, window wells can be drained by digging out the soil at the window well area to a depth of 18 to 24 inches. Lay a solid 4-inch PVC pipe with an elbow and grill drain with the top end of the pipe 4 to 5 inches below the window opening or sill (*see Figure 10*). Install the areaway; if the discharge pipe obstructs the areaway, cut out a section of the areaway to allow for a proper fit around the pipe. Lay crushed stone in the areaway to 1 or 2 inches below the top of the drain grill cap. Continue to lay drain pipe on a downgrade flow or into a dry well.

PVC pipe and drain cap

A HOUSE WITH A STONE FOUNDATION

When installing a French drain system in a basement that has a stone foundation, cut the concrete floor approximately 14 inches, starting from the stone foundation on the sides that have a water problem. Dig the soil in the trench on a slope away from the base of the foundation wall. As stated earlier, always start a French drain system in the sump pump area and install the chamber and sump pump first, to handle water that seeps into the trench. If water should enter faster than expected, 4-inch perforated pipe can be laid in the trench and covered with stones to subgrade. This is the best way to secure the existing condition overnight in case it rains.

If the stone foundation wall leaks, lay a strip of plywood on edge over the stone in the trench, in the areas of leakage, and pour concrete in the trench against the plywood. When finishing the concrete, run a concrete edger against the plywood; remove the plywood the same day. The slot will provide a weep space for water

Weep space

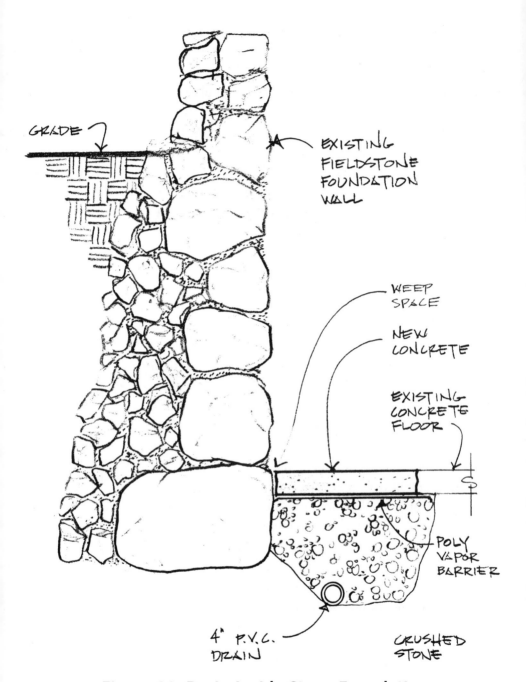

Figure 11: Drain Inside Stone Foundation

to enter the system (*see Figure 11*). Eliminate the weep space where soil is very wet and sandy; it will provide an undesired entry for sand into the basement.

Do not proceed with a French drain system if a test hole reveals a high water table. It is best to wait for drier conditions.

INSTALLING A DRAIN SYSTEM OUTSIDE OF FOUNDATION WALLS

To waterproof without disturbing a finished basement, an outside drainage system is preferred. First, dig outside on all sides of the foundation to 12 inches below the existing concrete floor. If the basement is concrete block, cover the foundation wall with a roofing plastic cement. Then lay two sheets of polyethylene over the plastered surface. To prevent the polyethylene from sliding down when you backfill, secure it with strapping nailed to the wall. Lay continuous perforated 4-inch PVC pipe, which will enter two sump pump holding chambers inside the basement, perhaps in a furnace room or utility area.

Place panels of 2-inch-thick, 36-inch-wide rigid insulation against the outside foundation walls. If the foundation is concrete block, it is important to insulate to at least 48 inches below grade; in very cold climates, insulating all the way down to the footings is even better. Backfill the entire trench with stone to grade, to prevent frost from exerting damaging lateral stress on the block foundation (*see Figure 12*).

Insulation

In the event of a power failure, it is best to have an emergency generator that will continue to run the sump pumps, especially when you have a finished basement to protect.

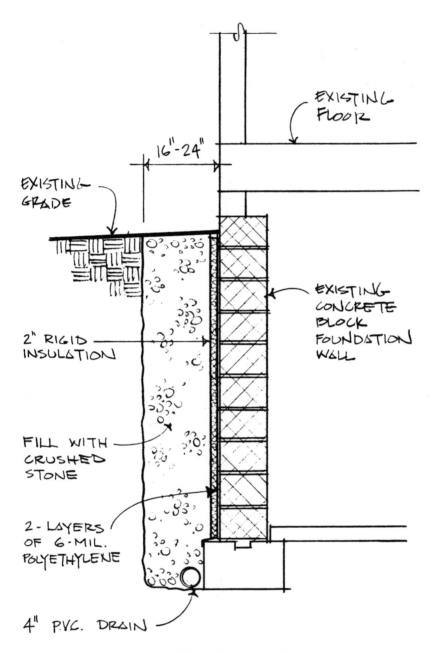

**Figure 12: The Outside System—
Concrete Block Foundation**

A CONCRETE BLOCK FOUNDATION WALL

Water problems are often a result of downspout discharge water, which enters the basement through the mortar joints of concrete block foundation walls. Water moves through the porous block from one chamber to another, raising the water table inside the block chambers. Eventually, water will seep onto the floor.

A method of releasing water from the cells of concrete block is to cut slots in the concrete footing that intersect the block. These slots allow water to flow freely out of the block and into the French drain system.

Every house with a basement should have gutters and downspouts that discharge water away from the foundation on a downgrade flow. Improper water drainage will cause dampness in the basement or will cause water to enter the basement from under the foundation wall and footing. This water will be noticed first where the basement floor abuts the foundation wall.

Downspout discharge

DAMP-PROOF BELOW-GRADE WALLS

All concrete foundation exterior walls in homes should be damp-proofed below grade with a waterproofing membrane system, emulsified asphalt compounds, or any other suitable product (*see Figure 13*). This will be very effective in stopping water from flowing through cracks in the wall and wall ties. Cover all waterproofing walls with a 2-inch-thick rigid insulation to grade.

Damp-proofing systems

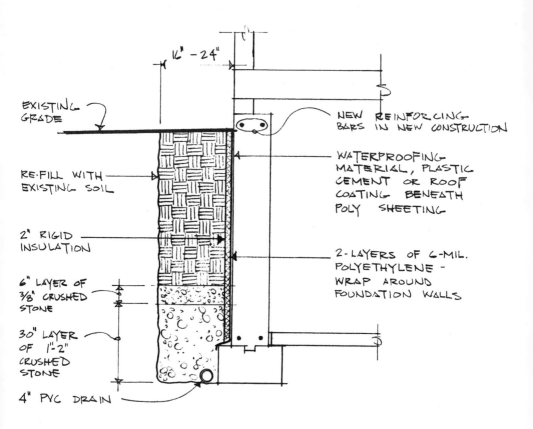

Figure 13: Damp-Proof Foundation Walls

HYDROSTATIC PRESSURE UNDER A CONCRETE FLOOR WHERE HEAD ROOM IS AVAILABLE

In a wet basement where hydrostatic pressure has raised the concrete floor, the best method of correction is to use the following procedures:

Raise floor when feasible

1. Cut the concrete floor next to the foundation wall approximately 12 inches in width on all sides, leaving uncut sections of floor as wall supports as shown in *Figure 2*.
2. Dig a trench 7 inches deep.
3. Install two large sump pump chambers 7 inches above the old existing floor.
4. Using a jackhammer with a bull point pin, punch holes into the existing concrete floor approximately 48 inches apart. This releases hydrostatic pressure under the floor and allows the water to flow up into the stones and pipe system leading to the holding chambers.
5. Lay perforated pipe in the trench, and cover the entire floor area over the pipes in the trench with 3 inches of crushed stone.
6. Lay a 6-mil polyethylene sheet over the stones, and pour a new 4-inch-thick concrete floor with a 4-inch slump and a compressive strength of no less than 3500 psi (*see Figure 14*).

Do not underestimate hydrostatic pressure. I have seen houses where water pressure has raised a 12-inch-thick concrete floor 3 to 5 inches, along with the main carrying timber and Lally columns with footings.

Water is powerful

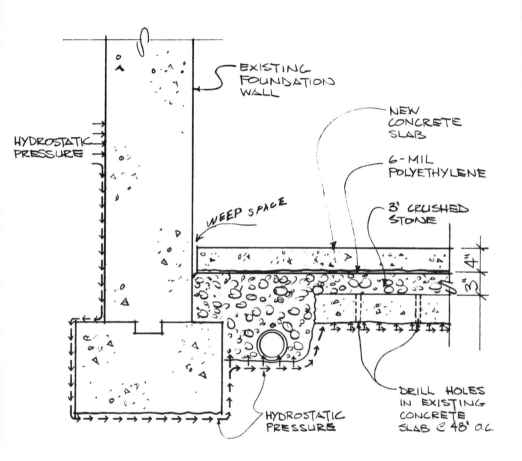

**Figure 14: Hydrostatic Pressure—
New Floor Poured Over Old**

HYDROSTATIC PRESSURE UNDER A CONCRETE FLOOR WITH SANDY SOIL AND NO HEAD ROOM

Sometimes soil under the concrete floor is sandy, having a fast water recovery that raises and cracks the concrete floor. Do not attempt any repairs while the ground is wet. When the soil is dry and the weather forecast is for fair conditions, use the following procedures:

When floor can't be raised

1. Cut out the entire concrete floor and remove 4 to 5 inches of soil throughout the entire sub-floor area.
2. Dig next to the foundation wall, at a slope of 45 degrees away from the wall, a trench 14 inches wide and 12 inches deep from the top of the oncoming floor.
3. Lay plastic pipe in the bottom of the trench with the holes up to discharge into the two holding chambers or the gravity flow system if conditions will allow.
4. Place 4 to 5 inches of crushed stone over the pipes and entire floor to subgrade.
5. Lay a sheet of 6-mil polyethylene over the entire floor, and pour the new concrete floor.

A SINGLE SUMP PUMP INSTALLATION

A sump pump is installed by cutting the concrete floor and digging a 27-inch-deep hole in a corner of the basement. Cut out two 4x4-inch or 4¼x4¼-inch holes on the two sides of the holding chamber (*see Figures 15-16*). Lay 2 inches of stone at the bottom of the hole. Place the chamber upright on the stones and fit galvanized wire mesh over the cutout holes to prevent the stones from entering the chamber. Fill around this chamber with stone to 4 inch-

Similar to double pumps

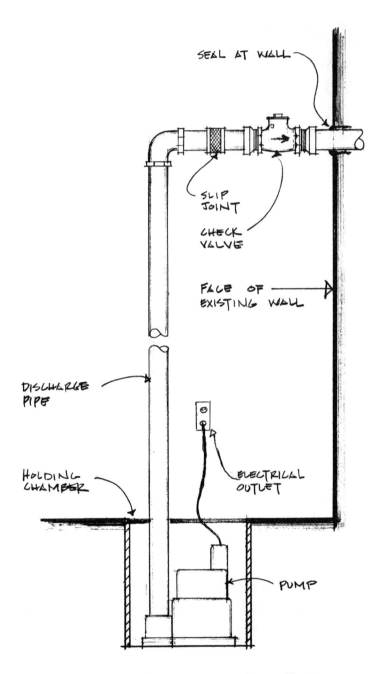

Figure 15: Sump Pump Installation

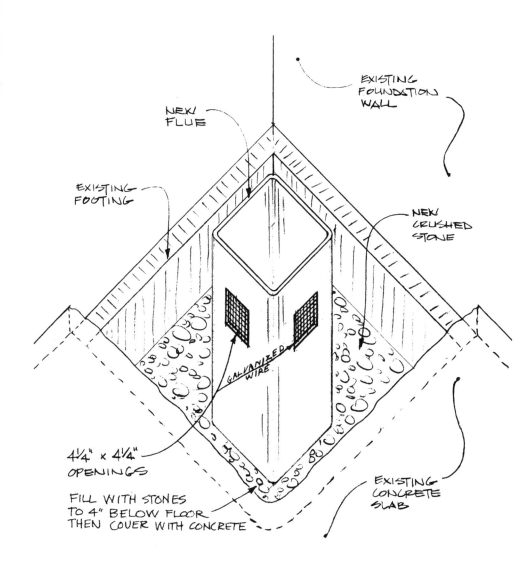

Figure 16: Single Sump Pump Chamber

es below the floor, then place concrete at the bottom of the holding chamber over the stones (precast patio block works well), and install the sump pump on top of the slab. If a French drain system is installed later, the perforated pipes can enter the cutouts in the holding chambers.

Where dwellings are close together and the pitch of the land will not allow the water to leave the sump pump without causing damage to abutting properties, it is best to discharge water into the nearest storm drain (if this is legal) or to install a dry well (see page 41 for details).

A HOUSE BUILT ON WET LAND

Anticipate trouble

When building a house where the water table is high, you can take certain steps to avoid water problems. First, a basement concrete floor grade should be established. Mark a floor grade chalk line on the foundation wall 5 inches above the footing. Install sump pump holding chambers with the cutouts in a corner of the basement floor. Lay 4-inch plastic piping continuous at the base of the footing to return into the chambers. Lay 3 inches of crushed stone over the entire floor area and cover the stone with polyethylene sheeting. Wrap the bases of Lally columns with styrofoam to separate the bases from the concrete floor to prevent cracking. Then pour a new concrete floor over the entire area.

A HOUSE BUILT ON A DOWNGRADE SLOPE

Waterproofing materials

A house built on a downgrade slope will almost always have a water problem. To control the situation while the concrete foundation walls are still exposed,

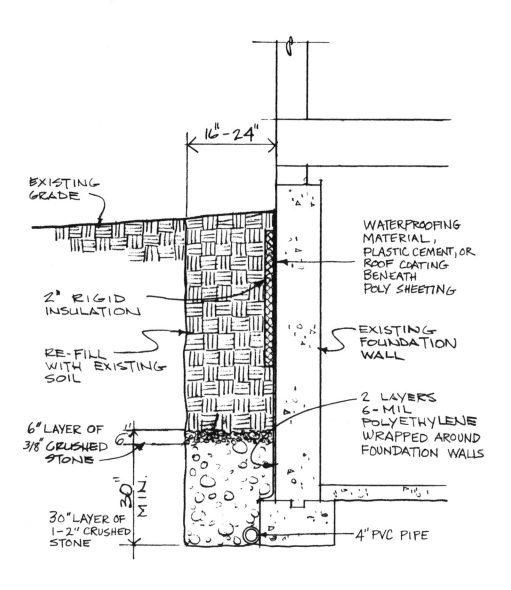

Figure 17: The Outside System—Concrete Foundation

apply a waterproofing material that will fill the concrete pores (see "Supplies" on page 68). If such a product is not available, apply a plastic cement or roof coating with a plastering trowel and wrap the foundation wall with two layers of 6-mil polyethylene over the plastered surface.

After either of the above applications, lay 4-inch plastic perforated pipe with the holes up at the base of the footing, continuing around the three sides of the upper grade of the house. These drainage pipes will continue on to a gravity flow. Backfill over the pipe system with at least 30 inches of 1- or 2-inch crushed stone and lay 6 inches of ⅜-inch crushed stone over this system to keep the oncoming topsoil from filtering through the larger stone and entering into the pipe system. Next, lay rigid insulation against the foundation and continue backfilling to grade (*see Figure 17*).

Downspouts should discharge separately

Downspouts for the house, which often contain leaves and other debris, should always have their own solid-pipe drainage systems for discharging water away from the building.

ESTABLISH HOUSE GRADE

Avoid lowest point

When building a house on a downgrade slope, where a house foundation is built much lower than the existing road grade, you should first consider elevating the house foundation as much as necessary to overcome possible water problems. Grade from the house to the road with a reverse grade, forming a valley so water can flow away from the house. To avoid problems with water, lay perforated pipe along the perimeter at the base of the footing on the exterior of the foundation. Then backfill with crushed stone approximately 30 to 36 inches deep and follow *Figure 17*.

INSTALLING DRAINS IN FRONT OF GARAGE DOORS

Water problems often exist when a garage is built under the house and the driveway has a downgrade slope. Water naturally follows the surface of the driveway and enters the garage. A method of correcting the water entry is to dig a trench to a depth of 24 to 36 inches and approximately 16 inches wide across the entire frontage of the garage opening. Lay crushed stone in the trench and pour a 4-inch concrete slab over the stone before installing a prefabricated channel slope system with grates. These trough sections are laid in front of the garage doors with pipes to flow on a downgrade left or right if possible (*see Figures 18-19*). If gravity drainage is not possible, you will need to install a lift chamber or a dry well.

Another method for collecting water in this situation is to install a concrete block manhole with an open grate top and a 6-inch plastic discharge pipe 16 inches below the top of the grate down to the pipe invert. This concrete manhole should be placed beneath the center of either a two-door garage or a single-door garage.

A mini catch basin and manholes are optional to use in *Figure 20*.

These precast units can save time and be more adaptable to some doors and extended around the side of the building to discharge at the back. Use a pump or a gravity flow. The mini catch basin also can be used as a lift chamber or dry well.

Channel slope system

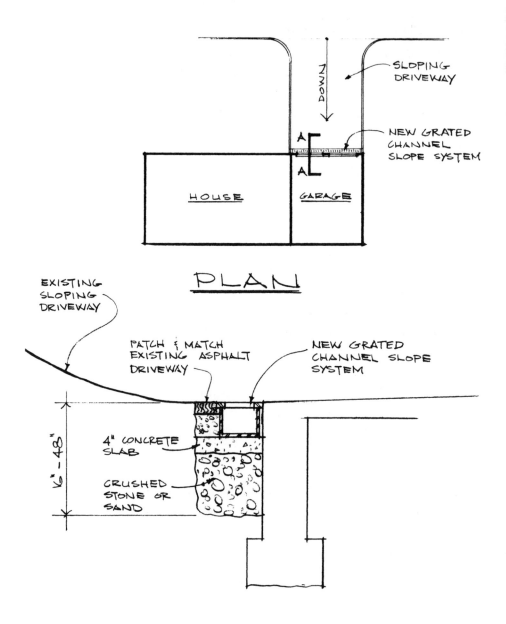

Figure 18: Driveway Sloping Down to Garage

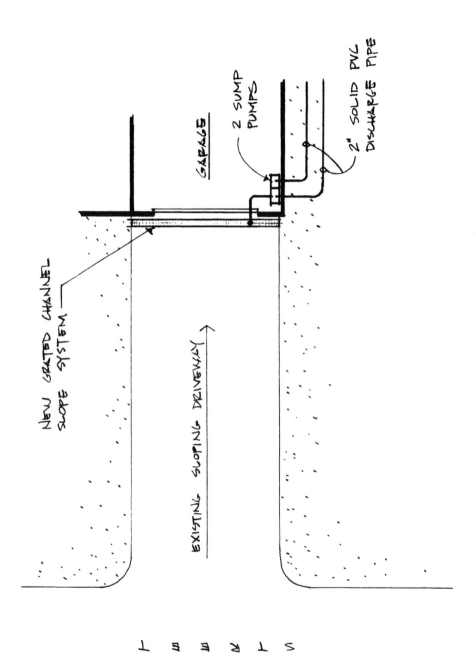

Figure 19: New Grated Channel Slope System

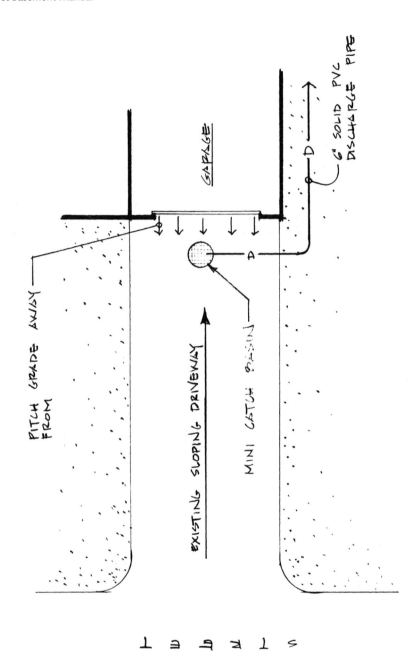

Figure 20: Channel Slope System With Optional Mini Catch Basin

A SUMP PUMP LIFT CHAMBER

In wet locations without the advantage of gravity-aided water discharge, a complete block chamber may be constructed at the lower discharge end of an existing drainage system. A common size is 40x40 inches by 4 feet deep, with a concrete slab at the bottom. Then set 8x8x16-inch concrete block in masonry mortar around the inside walls of the chamber, and fill the block cells solid with mortar as you build the chamber. (This custom-built concrete block chamber can be made as large or as small as required. Also, you may install sump pumps as necessary to dewater the chamber.) See *Figure 21* for details on building the chamber.

The sump pump is installed on the floor of the chamber. The discharge water will be pumped through a plastic pipe with a check valve installed on the horizontal section at the top of the elbow leaving the chamber. The discharge pipe will leave the chamber on a downgrade slope so that water will not remain in the pipes and freeze. If the discharge water is pumped on an upgrade and not below the frost line, a small hole can be drilled in the center of the flap in the check valve to allow the water to return slowly into the sump pump chamber. This procedure will prevent water from freezing in the pipes. The cover for the chamber should be solid to keep out rainwater. Make it as strong and secure as necessary to prevent tampering or accidents.

The lift chamber can also be used to contain a new source of water that might result when a new development abutting your land is graded to a higher elevation. First, dig a trench the length of the standing water, approximately 16 to 20 inches deep and 16 inches wide. Pitch the trench to the left or right to assist

Custom-built well

Create good drainage

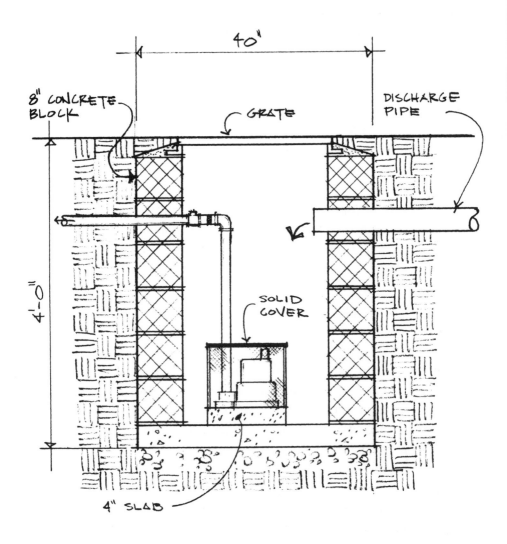

Figure 21: Sump Pump Lift Chamber

drainage toward a gravity flow or a lift chamber. After 4-inch perforated plastic pipe has been laid in the bottom of the trench, fill the trench to existing grade with crushed stone.

A DRY WELL CHAMBER

A dry well chamber may be built when discharge water cannot be pumped into the back yard or where a storm drain does not exist. Locate the dry well chamber away from the house on a downgrade slope so the water will not re-enter the basement and the discharge pipe will have a pitch to the chamber. First, dig a hole approximately 6x6 feet and 6 feet deep. Place 4 to 5 inches of crushed stone into the hole.

Drains on its own

Build a concrete block chamber by laying concrete block over the stones and filling the block cells with crushed stone to prevent the block from shifting (*see Figure 22*). Next, lay a heavy plywood cover over the top course of block resting 3 inches on all sides. Then build a form 5 inches high over the plywood cover. Lay #4 rebar 12 inches on center each way, then pour 3500 to 4000 psi concrete. Finish by filling around the chamber, to subgrade, with crushed stone. Then cover with double polyethelene and, finally, existing soil to grade.

A HOUSE BUILT ON
A CONCRETE SLAB

A house built entirely on a concrete slab may take in water because of improper landscaping, enabling water to flow toward the house slab. This problem can be corrected by digging outside next to the leaking side

Water in house

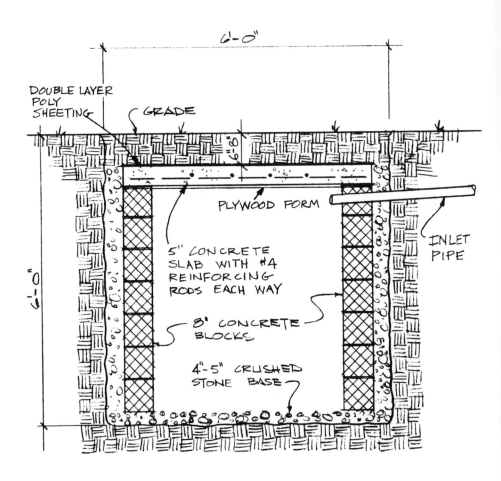

Figure 22: Dry Well Chamber

of the house slab to a depth of 12 to 16 inches and approximately 16 inches in width, continuing to a downgrade slope for gravity flow. If a gravity flow does not exist, a dry well chamber or a lift chamber can be installed to dissipate the water or to pump the water away to another area. A 4-inch pipe is laid at the bottom of the trench that will pitch to discharge. Fill the entire trench with stone to grade (*see Figure 23*).

OUTSIDE SUMP PUMP CHAMBER

When a French drain installation in a finished basement does not allow space for a sump pump, what is needed is an outside concrete lift chamber that will accommodate a sump pump to keep water out of the system.

Insufficient space

Dig on the outside of the house foundation wall to a depth of 30 to 36 inches below the top of the existing basement floor. Pour a concrete slab and lay the precast wall pipe sections in cement mortar. Make sure the pipes that flow into the chamber are at least 24 inches below the basement floor. After all of the wall pipe sections are laid in cement mortar to grade, install a solid cover on top of the chamber.

One of the local precast concrete producers in my area has sets of forms to pour a mini catch basin and manhole in sections. The first section contains a concrete floor and a chamber section with knockout openings. If needed, other sections also have knockout sections and come in different lengths. This product may be available in your area.

Form for mini catch basin

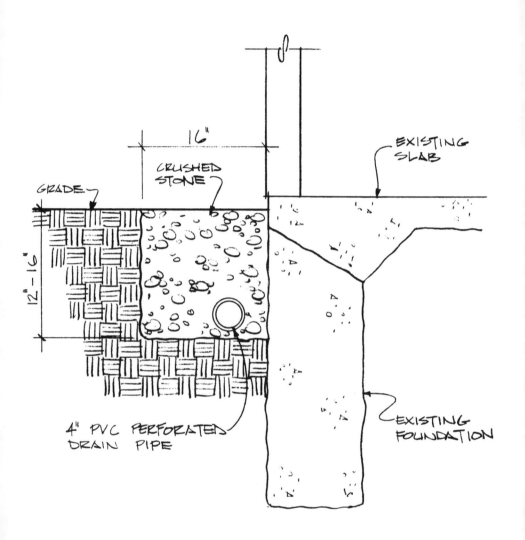

Figure 23: Outside Drain System—House on Slab

CONTROLLING DOWNSPOUT WATER

Sometimes water problems are created simply by the water discharging from downspouts next to the house foundation wall. In this case, dig a trench from the base of the downspouts 10 to 12 inches deep, continuous on a downgrade slope. Next, install a plastic downspout connector onto a 4-inch solid pipe and elbow to divert the water away from the house on a downgrade flow. This procedure may be all that is necessary to correct your water problem (*see Figures 24 and 25*).

If a gravity flow is not possible, a dry well as illustrated in *Figure 22* should be built no closer than 15 feet away from the house (*see Figure 26*).

USING A BLOW PIPE TO EXCAVATE UNDER OR AROUND OBSTRUCTIONS

The blow pipe is an improvised tool that can be very useful in removing soil from difficult places, such as from underneath the concrete floor of a finished basement bathroom. (*See Figure 27 for a detail of a blow pipe.*) Note that additional lengths of blow pipe ends can be attached in order to reach the desired depth. The most popular lengths are 24 inches, 48 inches, and 72 inches. A small air compressor to power the blow pipe can be purchased or rented.

Compressed air

When installing a French drain system in a basement where an oil tank or furnace is an obstacle, cut the concrete floor and trench up to each end of the tank. (It is important to dress with protective clothing, including a hood, while doing this procedure.) Turn the ball valve

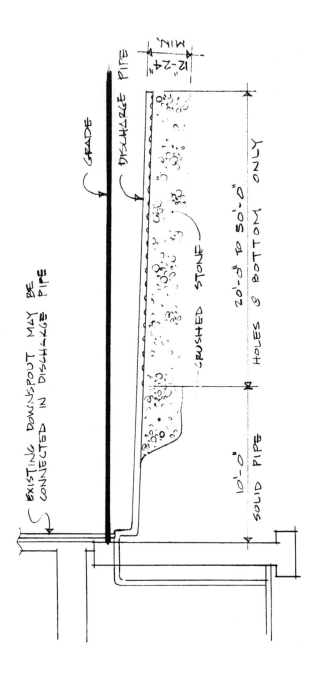

Figure 24: Downspout Water Diversion

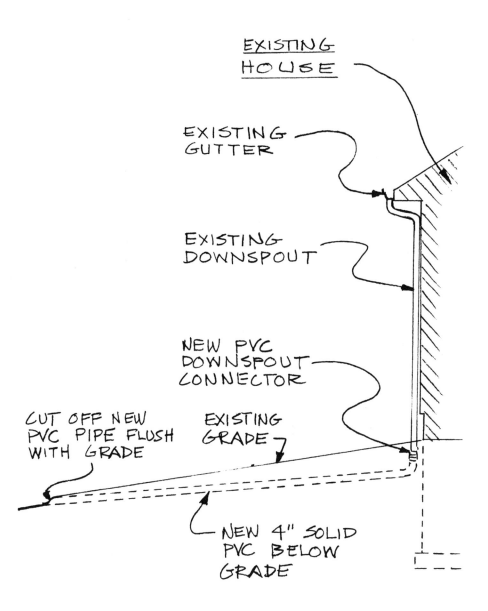

Figure 25: Downspout Extension

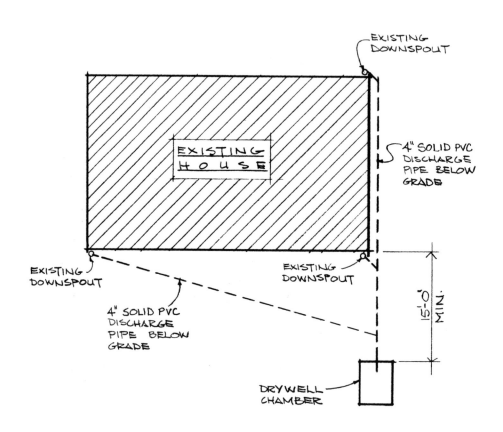

Figure 26: Downspout Extension to Dry Well

of the blow pipe to allow the compressed air to blow full blast through the pipe as it is being thrust forward and backward against the soil that is being blown back. Continue on each end of the tank and remove loose soil with a hoe and shovel until the hole is the same size as the existing trench. Then lay the perforated pipe continuous through the tunnel. Backfill the entire system with stone using the handle of a shovel to poke stone into the tunnel and around the pipe. Complete the project by pouring concrete over the stone-filled trench (*see Figure 28*).

There are some instances where the blow pipe procedure cannot be used because the soil is composed mainly of clay or it contains too many large stones. In these cases, it is better to continue the trench around the oil tank or furnace (*see Figure 29*), always providing a weep space along the uncut section of the tank base to collect any water coming up from the rear of the tank.

Around if not under

LEDGE ROCK IN THE BASEMENT

In the northeastern United States, where I live, ledge rock is commonly encountered when basements are excavated. Water may flow through or around ledge rock, contributing to moisture problems. Because of the difficulty of removing ledge rock, builders of many older basements poured around or barely over ledge rock outcroppings. To correct such situations, take the following measures:

Hard to remove

1. Establish a sump pump hole location
2. Dig a trench from the sump pump hole to sections of the wet basement by cutting the floor next to the

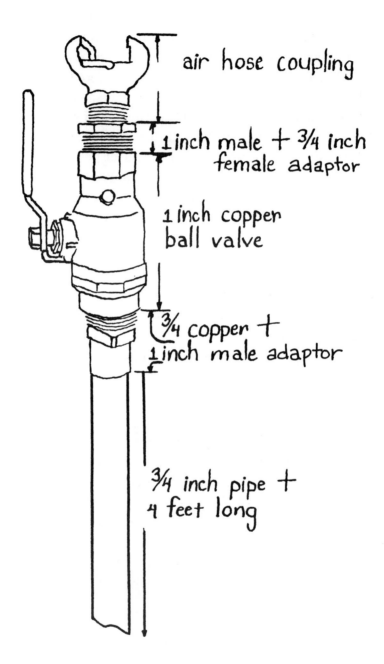

Figure 27: The Blow Pipe

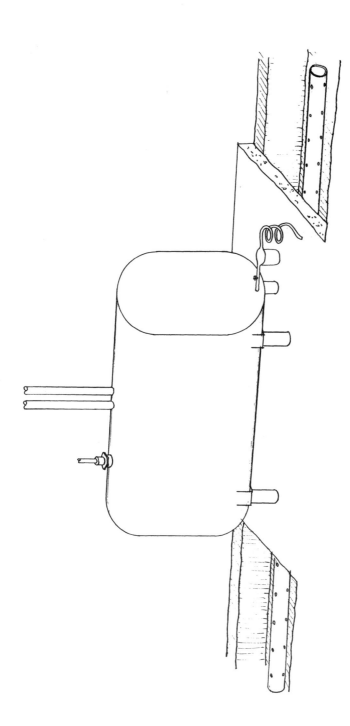

Figure 28: Blow Pipe Excavation Under Tank or Other Obstruction

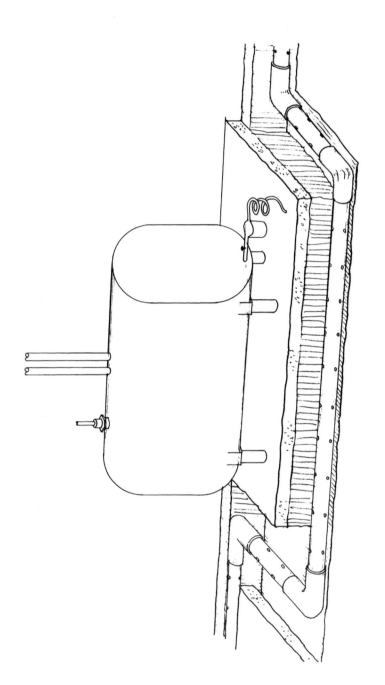

Figure 29: Excavation Around Tank or Other Obstruction (when you cannot blow through an obstacle next to the house foundation under the floor)

ledge rock and digging a 12-inch-deep trench on all wet sides.
3. Lay 4-inch perforated pipe and crushed stone, providing a weep space (a very important feature) to collect the flow of water from the ledge that will enter the French drain system.
4. Fill the trench with concrete, removing the weep space material at the same time (*see Figure 30*).

A HOUSE FOUNDATION WALL WITH NO FOOTING

Keeping in mind all procedures connected with the installation of the French drain system, special attention should be given to leaving enough contacts to secure the house foundation wall. If the foundation wall is not 12 inches below the existing concrete floor, the remaining depth of the trench should be dug at a 45-degree angle away from the wall. Next, lay the perforated pipe and the stone and continue by resurfacing the trench (*see Figure 31*).

Angle away from the wall

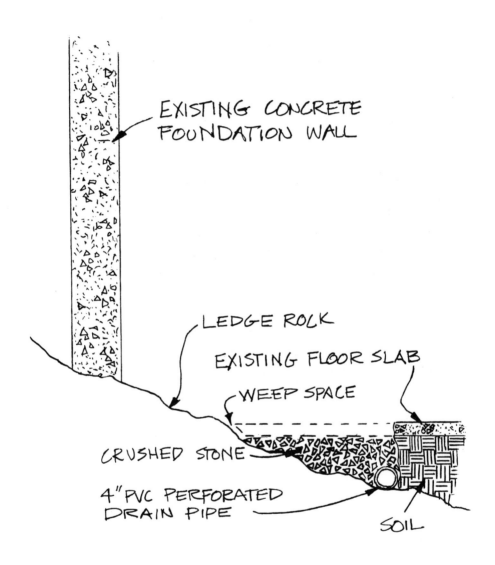

Figure 30: Exposed Ledge Rock

Wet Basement Manual

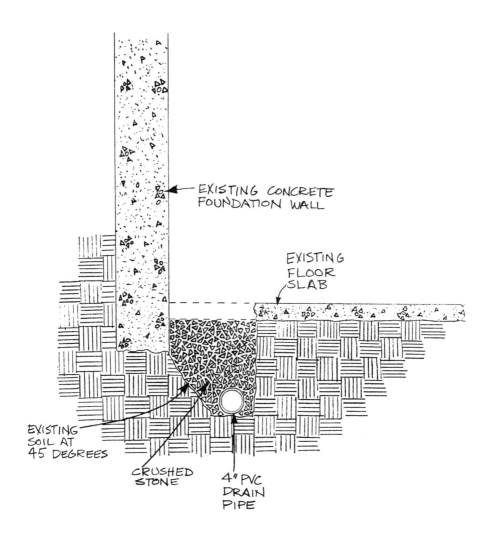

Figure 31: French Drain for Footingless Wall

SUPPLEMENTARY INFORMATION

High Humidity in Basement 57
A Cold Wall 58
Leaking Cracks—Outside Repair 58
Leaking Cracks—Inside Repair 59
Osmosis 59
Ferric Oxide or Hematite 60
Radon is an Invisible Threat 62
Shortcuts to Trouble 67
Tools 67
Supplies 68

HIGH HUMIDITY IN BASEMENT

The most practical way to prevent damp basement walls is to apply a damp-proofer such as Xypex to exterior foundation walls while the building is under construction. Follow the manufacturer's instructions and finish with a layer of 2-inch insulation board to grade.

If the builders failed to provide effective damp-proofing, a damp basement in an existing house can be corrected by digging outside the foundation walls to the top of the footing. Clean the wall surface and apply a plastic roofing cement up to the grade line. Cover the walls with two layers of 6-mil polyethylene and place 2-inch-thick insulation board over the polyethylene. If the problem is only dampness, finish the work by backfilling with the same soil you removed. If a major water threat and flooding are likely, install perforated pipe and stones as discussed for block foundation walls.

For serious humidity and dampness

If these procedures are not possible, the following suggestions could help: (1) Keep open all basement windows for cross-ventilation to help reduce dampness. (2) Purchase a heavy-duty dehumidifier, which will help enormously in keeping the basement dry. The water collected by the dehumidifier can be drained into the existing system by attaching a hose to the back of the dehumidifier (the water-holding pan) and allowing the water to drain into a sleeve pipe pre-installed to receive discharge water.

A COLD WALL

Condensation

Especially during a series of cold spells in winter months, one or two basement walls will receive less sunshine. Water may condense on these colder walls and run down the surface. If this condition is noticed while waterproofing the basement, install a weep space on the sides that have wet walls. Another method of correcting the dampness of cold walls is to dig on the outside of the house foundation wall and install 2-inch rigid insulation board to a depth of at least 48 inches. This will prevent coldness from penetrating the concrete walls.

LEAKING CRACKS— OUTSIDE REPAIR

Plastic roofing— cement and polyethylene

To prevent water from entering through cracks in the foundation walls, consider the following procedures for repairing from the outside:
1. Dig from the outside of the foundation wall down to the top of the footing.
2. Dry the concrete surface 6 inches from each side of the crack with a blow torch or heat gun.
3. Brush the areas clean and apply a ½-inch-thick layer of plastic roofing cement over the crack, 6 inches on each side, from the top of the concrete footing to the top of the surface grade.
4. Lay a 6-mil polyethylene sheet (16 inches in width) over the crack, starting from the bottom of the wall, including the top of the footing, and continuing to the top of the foundation wall. Be sure to secure the polyethylene sheet by nailing it, together with a strip of

wood, onto the house. This procedure is necessary to prevent the polyethylene sheet from sliding down and disturbing the tarred crack during trench backfilling.

LEAKING CRACKS—INSIDE REPAIR

Patching inside

Over the years we have developed a very successful method for repairing a cracked foundation wall, leaking wall ties, and leaks around pipes by applying a flexible rubberized patching material. The steps for applying this material must be followed carefully. First, the area must be absolutely dry and warm (at least 50 degrees F) before it can be repaired. Remove all laitance, efflorescence, paint, whitewash, and foreign substances within 3 inches of either side of the crack. After stirring well, apply the patching material over the prepared surface with a brush, working from the bottom of the crack up, covering the entire prepared surface. Follow the manufacturer's directions to complete the work. For more information, see "Supplies" on page 68.

OSMOSIS

Temperature difference

Sometimes water on the basement floor, especially in a deep basement, is the result of the diffusion of fluid through a semi-permeable membrane (the concrete floor), caused by the temperature difference between the room and the concrete floor. These differing temperatures can form a vacuum and draw water onto the surface of the floor. Very helpful measures in these circumstances include the creation of cross-ventilation and the use of a heavy-duty dehumidifier.

FERRIC OXIDE, OR HEMATITE

Prevent clogging

Pump-clogging ferric oxide, or hematite, is sometimes encountered when installing a French drain system. This substance, which has a gritty texture, can be identified by its red, brown, black, orange, or yellow color.

While installing a French drain system, it is important to install a 12x12x24-inch (or larger) sump pump holding chamber in each inside corner of the basement. Ferric oxide will be discharged into these holding chambers.

As iron-contaminated colored water flows into the French drain system, fine particles of ferric oxide will cling to the stones and pipe perforations, restricting water flow into the holding chambers. To allow these particles to flow more freely into the system, you need to double the number of holes in the 4-inch-diameter plastic pipe around the perimeter of the basement (*see Figure 32*).

The holes should be drilled on the top and sides of the pipe but not on the bottom. After the pipe is laid in the trench with the holes up, place 1½ inches of clean, crushed stone over the pipes to the subgrade of the concrete floor. Using larger stones, which form larger voids, enables the water to flow faster, thus helping to prevent ferric oxide from restricting the flow.

Flush each spring

To maintain this system, you should flush out the pipes and clean out the holding chambers in the late spring of each year. This can be accomplished using a garden hose with a brass nozzle, progressing slowly through each pipe section. Flush out the system while the ferric oxide is still soft in the pipes.

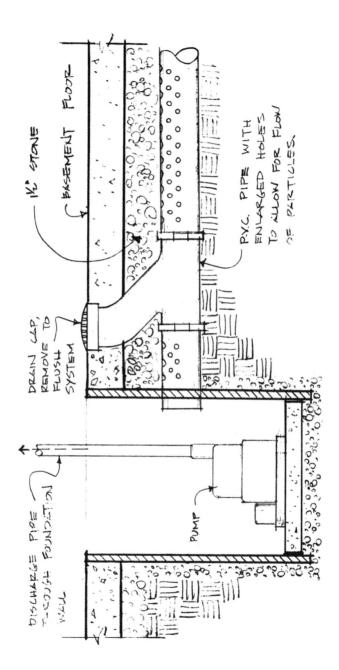

Figure 32. Ferric Oxide Stain Removal System

RADON IS AN INVISIBLE THREAT

Radon, an odorless and invisible radioactive gas, is a known carcinogen. Continuous exposure to high levels of radon gas can cause lung cancer. The Environmental Protection Agency (EPA) recommends testing all homes for the presence of radon; if the radon level is at least 4 picocuries (pCi/L), reduction measures should taken.* Radon is most likely to be present in a basement, since the sources of this gas include earth and rock beneath a home and well water.

After a test determines that radon is present, venting pipes and fans can be adapted to an existing French drain system (*see Figures 33-36*). These will be effective in removing radon from under the floor and through the sump pump holding chamber. In large basements, pipes can be laid across the concrete sub-floor and connected onto the existing pipes of the French drain or perimeter pipes.

In new construction, lay a 6-mil sheet of polyethylene over the sub-floor and pipes, and install a vertical stack vent with a small fan. Seal all of the gross openings in basement expansion joints around pipes and sump-pump covers.

---

*For more information on radon, visit the Environmental Protection Agency's Web site at http://www.epa.gov/iaq/radon, or call the National Radon Hotline (800-767-7236).

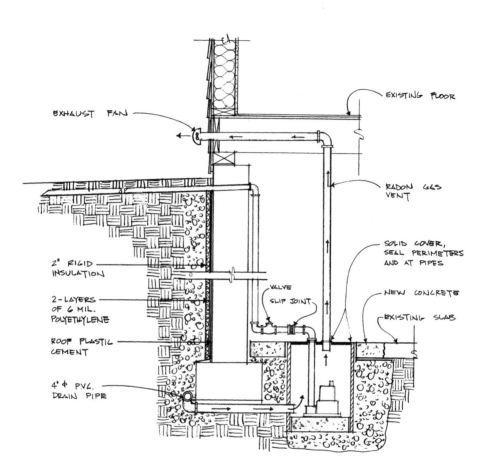

Figure 33: Water and Radon Removal

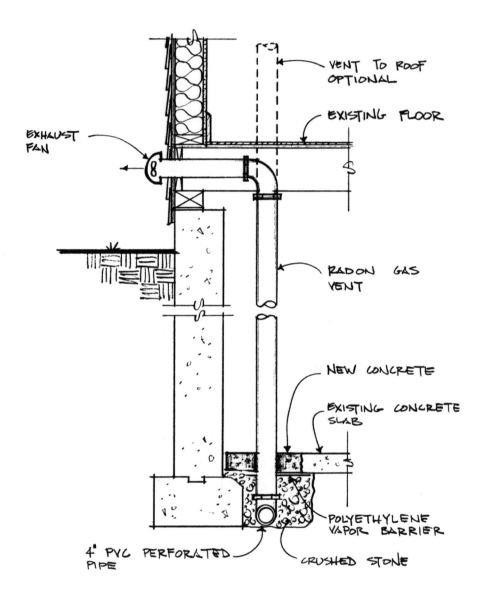

Figure 34: Radon Gas Removal System—Dry Basement

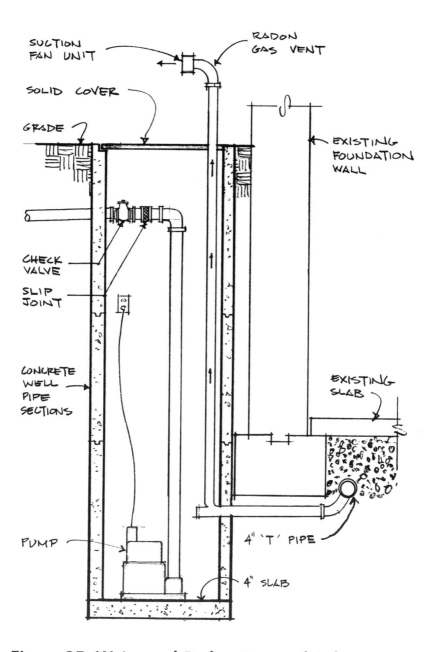

Figure 35: Water and Radon Removal (when a sump pump is installed in an outside concrete chamber)

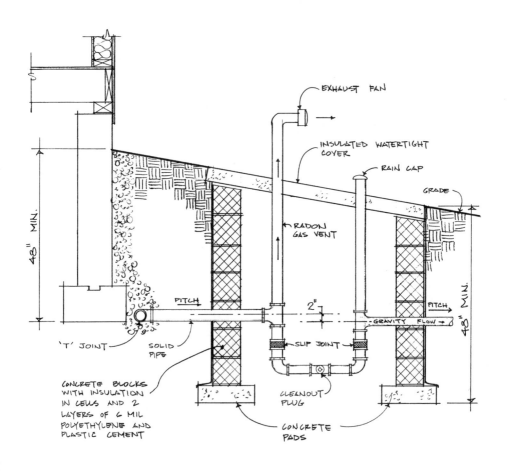

Figure 36: Water and Radon Removal—Outside System

SHORTCUTS TO TROUBLE

Some contractors believe they can avoid water problems simply by laying 6 to 12 inches of crushed stone under the concrete floor and (1) not laying perforated pipe around the perimeter of the basement; (2) not providing a sump pump and holding chamber; and (3) not providing a discharge gravity flow pipe from the system. Without the above precautions, water can enter under the floor, creating hydrostatic pressure and causing the floor to rise. To avoid wet basement headaches and expenses, don't take shortcuts. Crushed stone alone is not enough to prevent serious water problems.

TOOLS

The following tools are suggested for use with this manual:

stone drill	abrasive masonry blade
hammer drill	pipe wrench
pointed shovel	wheelbarrow
garden hoe	5-gallon pails
jackhammer	concrete wood float
push broom	concrete finishing trowel
power saw	edger
clay spade	concrete cutting blade

An air compressor for the blow pipe and other tools can be rented. Sizes from 85 to 185 cubic feet per minute are available. Consult with your supplier to match the size of the air compressor with the tools you are using.

SUPPLIES

Easy Sealment is a patented product developed by the author. It is a bonding agent and sealer useful for repairing leaking cracks and tie holes.

Xypex is a Canadian product consisting of portland cement, very fine treated silica sand, and active proprietary chemicals. When mixed with water and applied to concrete, a catalytic reaction seals the concrete against water penetration.